MATERIALS
ALL AROUND US

SEPARATING
MATERIALS

Robert Snedden

H www.heinemann.co.uk/library
Visit our website to find out more information about Heinemann Library books.

To order:
☎ Phone 44 (0) 1865 888066
▤ Send a fax to 44 (0) 1865 314091
▢ Visit the Heinemann Bookshop at www.heinemann.co.uk/library to browse our catalogue and order online.

First published in Great Britain by Heinemann Library,
Halley Court, Jordan Hill, Oxford OX2 8EJ
a division of Reed Educational and Professional Publishing Ltd.
Heinemann is a registered trademark of Reed Educational & Professional Publishing Ltd.

OXFORD MELBOURNE AUCKLAND
JOHANNESBURG BLANTYRE GABORONE
IBADAN PORTSMOUTH (NH) USA CHICAGO

Designed by Celia Floyd
Originated by Dot Gradations
Printed by Wing King Tong, Hong Kong

ISBN 0 431 12092 7 (hardback)   ISBN 0 431 12097 8 (paperback)
06 05 04 03 02   06 05 04 03 02
10 9 8 7 6 5 4 3 2   10 9 8 7 6 5 4 3 2 1

British Library Cataloguing in Publication Data

Snedden, Robert
   Separating materials. - (Material all around us)
   1.Matter - Properties - Juvenile literature 2.Materials science - Juvenile literature
   I.Title
   530.4

Acknowledgements
The Publishers would like to thank the following for permission to reproduce photographs: Andrew Lambert pp12, 13; Bruce Coleman Collection: Erik Bjurstrom p8; FLPA: M Rose p9, M J Thomas p28; Heinemann: Trevor Clifford p15; Hulton Getty: p19; Robert Harding Picture Library: p22; Science Photo Library: David Parker p5, Charles D Winters p7, Rosenfeld Images Ltd p14, Jerry Mason p26, Alex Bartel p29; Tony Stone Images: Phil Degginger p4, Rich Iwasaki p11, David Woodfall p18, Keith Wood p20

Cover photograph reproduced with permission of Science Photo Library

Every effort has been made to contact copyright holders of any material reproduced in this book. Any omissions will be rectified in subsequent printings if notice is given to the Publisher.

Any words appearing in the text in bold, **like this**, are explained in the Glossary.

Contents

Mixed materials

Some materials, such as stone and wood, can be used almost as they are, only needing to be shaped. However, most useful materials are almost always found with other materials that are not so useful. Others, such as plastics, have to be manufactured from **raw materials**.

Metals

Metals react with other **elements** to form **compounds** that are not useful to us. Before we can use the metals we have to find ways to separate them from the rocks, called **ores**, in which they are found.

In a foundry molten metal is separated from some of the impurities found with it.

Petroleum

Petroleum, or crude oil, is a complex **mixture** of many different oils and other compounds. Each of the oils has its own characteristics and its own uses, but before they can be put to use they have to be separated from each other. This is the job of oil refineries.

Water

The water that comes from your tap has many chemicals **dissolved** in it. The water companies have to ensure that any potentially harmful substances have been removed from it before it reaches you.

To separate the materials we need from the ones we don't need it helps to know how they have come together. This book is all about the different ways materials can be mixed together – and some of the methods used to 'unmix' them.

Desalination plants like this turn salt water into fresh water for drinking and irrigation.

Compounds and mixtures

All matter is made up of tiny **particles** called **atoms**. An **element** is a substance that is made up of just one kind of atom. Examples of elements include oxygen, a gas in the air we breathe, aluminium, a metal often used for making kitchen utensils, and calcium, found in chalk.

Elements can combine in a variety of ways to form millions of **compounds**. Examples of compounds include common salt, which contains atoms of sodium and chlorine, water, which has atoms of hydrogen and oxygen, and carbon dioxide, which has atoms of carbon and oxygen.

Atoms of different elements can join to form compounds. Carbon combines with oxygen to form carbon dioxide.

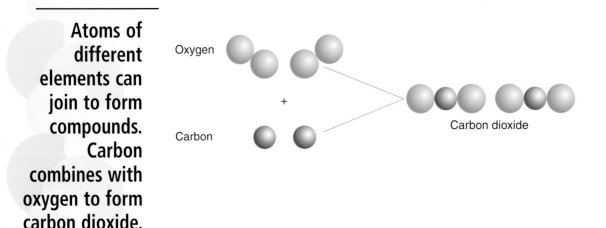

Oxygen

+

Carbon

Carbon dioxide

In proportion

In a compound the atoms always combine in the same proportions. Water always has two atoms of hydrogen to every one atom of oxygen, for example. Sometimes atoms of the same elements combine in different proportions to produce different compounds. Atoms of carbon and hydrogen can combine to form many compounds, such as natural gas, petrol and kerosene.

Change of states

Compounds may be solids, liquids or gases. Hydrogen and oxygen, which are both gases under normal conditions, combine to form water, which is a liquid at normal temperatures. Sodium, a soft metal, combines with chlorine, a greenish-yellow poisonous gas, to form sodium chloride (common salt), a hard, white solid.

A fizzy drink is a mixture of water, sugar, flavourings and carbon dioxide.

Mixtures

Many substances that contain atoms from more than one element are **mixtures**, not compounds. A mixture can be made up of different elements, or of different compounds, such as the water, sugar and carbon dioxide gas in a fizzy drink. Substances in a mixture do not react together to form new substances as they do in a compound and they can be separated by physical means such as **filtering** or distilling.

Solutions

A **solution** is a **mixture** that results when one or more substances – liquid, solid, or gas – are **dissolved** in another substance. The dissolved substances are called solutes and the substance in which they are dissolved is a called a **solvent**. Liquid solutions are most common. An example would be sugar (the solute) dissolved in water (the solvent). Gases and solids that dissolve in a liquid are said to be soluble.

The ability of a substance to dissolve in another is called its solubility. The solubility of most solids depends on the chemical properties of the solute and the solvent and on the temperature of the liquid solution. A hot solvent will dissolve more than a cold one. Eventually there comes a point when no more can be dissolved in the solution.

Fish use their gills to get the oxygen that is dissolved in the water.

Miscible liquids

Two liquids that can form a solution together are said to be miscible. Water and alcohol, for example, are completely miscible and will readily form a solution. Water and oil, on the other hand, do not mix well at all.

Solid solutions

A metal **alloy** – a combination of two or more metals such as the mixture of melted copper and zinc that forms brass – is an example of a solid solution. A tastier example would be an orange ice lolly made by freezing orange juice – a solution of fruit juice and sugar in water.

Gas solutions

The air we breathe is a gaseous solution, formed from a mixture of nitrogen and oxygen, plus smaller amounts of other gases such as carbon dioxide.

Solution caves are formed when water trickles down through tiny cracks in limestone and slowly dissolves the rock. They can take thousands of years to form.

Suspensions and emulsions

A **suspension** is a liquid or gas that contains small **particles** of another substance. Examples of suspensions include:

- a solid in a gas, such as particles of smoke drifting in the air;
- a liquid in a gas, such as fog, which is water **vapour** in the air;
- a solid in a liquid, such as the cloudiness seen in a muddy pond;

The foam used in fire extinguishers is a suspension of a gas, carbon dioxide, in a liquid.

- a gas in a liquid, such as the foam used in fire extinguishers;
- a liquid in a liquid, such as water-based paints.

Suspensions and solutions

A beam of sunlight shining into a room can be seen because tiny suspended particles of dust in the air reflect and scatter the light. This doesn't happen if light is shone through a **solution** because the particles are too small to scatter light. However, it is possible to separate out the particles in a suspension using a **filter** of the right size. But no filter is fine enough to separate out the **molecules** in a solution.

Emulsions

An emulsion is a type of suspension that is produced when one liquid is evenly distributed in another. It is not a solution – the two liquids do not **dissolve** in each other – tiny drops of one liquid are suspended in the other one. Milk is an emulsion of butterfat in water.

Smog, which can make life so unpleasant in large cities, is a suspension of pollutants in the water vapour in the air.

Sieves and filters

A **filter** is used to separate out unwanted substances from liquids or gases in a **mixture**. The filter has small holes that are large enough to let the liquid or gas **molecules** pass through but too small for the solid **particles**. The liquid that passes through the filter is called the **filtrate**. The solid that is left behind is called a **residue**.

Sieving is a way of separating larger particles from smaller ones.

A sieve is a type of filter that is used to separate out larger particles. When you drain cooked rice through a sieve you are filtering out the rice from the water. In this case you eat the residue! A tea strainer to trap tea leaves is another example of filtration – but in this instance it is the filtrate we are interested in.

Filters at work

Car engines use filters to remove impurities from air, lubricating oils and the fuel they use. For example, dry-paper filters on **carburettors** remove impurities from the air before it enters the engine. Water purifying filters use either **ceramic** or glass fibres to trap unwanted dirt and bugs. Air conditioners use filters made of fibre-glass or metal, coated with an adhesive, to remove dust and pollen from the air.

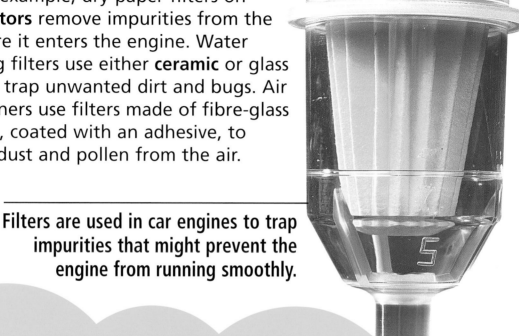

Filters are used in car engines to trap impurities that might prevent the engine from running smoothly.

Try it yourself

You will need
water
soil
two jars
coffee filter paper

1 Pour water into one of the jars and then stir a little soil into it to make a cloudy suspension.
2 Put the coffee filter paper into the top of the clean jar. Slowly and carefully pour the muddy water from the first jar through the filter paper, into the second jar.

The coffee filter will trap a lot of the soil (the residue). The water in the second jar (the filtrate) should look cleaner. However, it is not clean enough to drink. The water will probably still look fairly cloudy as you will not have been able to remove the finest soil particles - or the bacteria from the soil.

Clean water

Water treatment plants use a range of methods to make sure that the water we drink is clean and safe. The simplest way of cleaning up water is by filtration and settling. Settling simply means letting larger **particles** drift to the bottom where they can be removed. At the same time, waste materials that float on top of the water can be skimmed off. Half of the pollutants in the water can be removed by these simple methods.

Waste eaters

Useful **micro-organisms** consume waste material in special tanks. Solids and micro-organisms are then separated from the waste water in settling tanks. Disinfectants (such as chlorine) may be added to the water to kill any remaining disease-causing organisms.

A water treatment plant where harmful impurities are removed and the water is made safe to drink.

Filtering and flocculating

Chemicals from **pesticides** and cleaning materials are removed by **filtering** through specially treated carbon to remove organic materials and by **distillation** to remove salts. Flocculation is a process that involves adding a chemical to water that causes smaller particles to clump together and settle out.

Light cleaning

In 1996 American scientist Ashok Gadgil invented a water purifier that removes bacteria using ultraviolet light. The purifier did well against several germs, including E. coli bacteria, typhoid and cholera. A company in Bombay is making a solar-powered version that can be used in villages in India.

Some pollutants are extremely difficult to remove from water. If you pour some household cleaners or garden pesticides down the drain you are adding chemicals that might not be removed during treatment.

15

Evaporation and distillation

The **molecules** in a liquid are always on the move. The more energy they have, the faster they move. **Evaporation** is what happens when some molecules have enough energy to escape from the liquid as **vapour**. Energy in the form of heat can speed up evaporation. Puddles of rainwater disappear more rapidly when the sun shines. The increase in temperature increases the energy of the water molecules and so they escape at a faster rate.

The particles in a liquid are constantly on the move. Evaporation occurs when a particle has enough energy to escape from the surface of the liquid into the surrounding air.

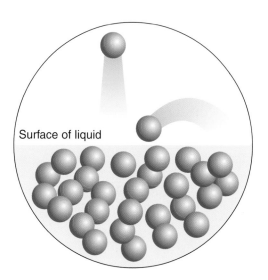

Surface of liquid

Volatility

Different substances evaporate at different rates. For example, alcohol will evaporate much faster than water at the same temperature. Substances that have fast rates of evaporation are said to be **volatile**.

Evaporation at work

Evaporating molecules remove heat as they escape so evaporation can be used for cooling. As perspiration evaporates from your skin it cools you down.

Distillation

Distillation separates a substance or a **mixture** of substances from a **solution** by vaporizing the liquid. It can be used to separate mixtures of liquids and usually involves heating the mixture and condensing the vapour that forms. The liquid collected in this way will be largely made up of the more volatile of the substances.

Distillation is used in oil refining to separate out the different parts of crude oil, or petroleum. This is called fractional distillation as the various gases and liquids that are obtained are called fractions.

Distillation can be used to separate liquids that boil at different temperatures.

Density and decanting

The **density** of a substance is found by dividing the amount of matter it contains by its **volume**, the amount of space it takes up. Liquids with different densities will separate out – as you can see with oil floating on water, for example – which allows the oil to be skimmed or poured off.

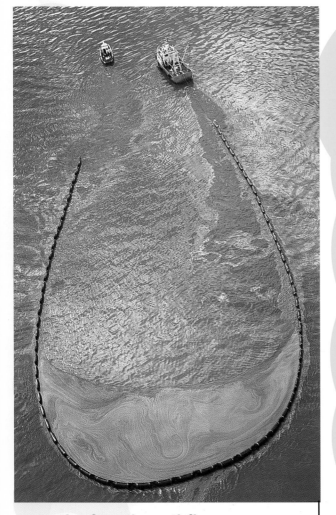

The fact that oil floats on water makes a spill slightly easier to deal with as some of the oil can be skimmed off the surface.

Try it yourself

You will need
a tall jar
syrup
glycerine
water with a drop or two
of food colouring
olive oil or sunflower oil

1 Carefully pour some syrup into the bottom of the jar.
2 Add some glycerine, carefully running it down the inside of the jar so it forms a layer on top of the syrup.
3 Do the same with the coloured water, followed by the oil.

If you've been careful you should have four layers of liquid with the most dense, syrup, at the bottom and the least dense, oil, at the top.

Decanting

Decanting is used to separate out substances of different densities from a **mixture**. If the **particles** in a liquid are dense enough they will sink to the bottom of a container and the liquid can be poured off leaving them behind. Cream can be removed from milk by allowing the less dense cream to rise to the top of the milk and then skimming it off.

Looking for gold

Gold prospectors used large, shallow pans to scoop up water, sand and gravel from a river where they thought there might be gold. They would swirl the mixture around in the pan and then pour out the muddy water, hoping to find some dense gold left behind! This method is called panning.

Washing and decanting to extract gold from rocks was hard work – but sometimes the rewards made it all worthwhile.

Oil refining

An oil refinery processes and refines hundreds of thousands of barrels of crude oil every day.

Crude oil, or petroleum, is mainly made up of a **mixture** of different **compounds** of hydrogen and carbon (hydrocarbons). The job of an oil refinery is to take this mixture and separate it out into useful products.

Fractional distillation

The first stage in oil refining separates crude oil into different parts, called fractions. The oil is pumped through pipes and heated to around 400°C. The mixture of hot gases rises through a steel tower called a fractional **distillation** column. As the gases rise, they cool and **condense**. Heavy fuel oils condense first and petrol and kerosene condense in the middle and upper sections of the tower. Gases such as methane and butane are collected at the top.

Cracking

Scientists have developed ways of converting less useful oil fractions into more valuable ones. Cracking converts larger hydrocarbon **molecules** into smaller ones.

The large molecules can be broken down using great heat and **pressure** or by the use of a catalyst to speed up the cracking process. A catalyst is a substance that changes the speed of a **chemical reaction** without being changed itself. Petrol produced by cracking is actually of better quality than that produced by fractional distillation.

Removing impurities

The most common impurities in oil fractions are sulphur compounds. These can damage machinery and are also a source of pollution when the fuels are burned. The fractions are mixed with hydrogen gas and then heated and exposed to a catalyst. The sulphur in the fractions combines with the hydrogen, forming hydrogen sulphide which is later removed by dissolving it in a suitable **solvent**.

Heavy fuel oils condense first followed by petrol and kerosene. At the top of the tower gases such as methane and butane are collected.

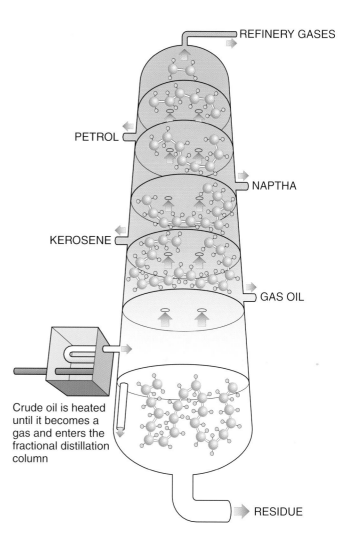

REFINERY GASES

PETROL

NAPTHA

KEROSENE

GAS OIL

Crude oil is heated until it becomes a gas and enters the fractional distillation column

RESIDUE

Ores and extraction

Most of the metals we use have to be separated from **ores**, the rocky material in which they are found. There are a wide variety of processes for getting metals from their ores.

Mineral dressing

Mineral dressing is done after the ore has been mined. The ore is crushed and then bubbled through water using gas. Chemicals or oils are added to the water to make the mineral **particles** stick to the bubbles. The minerals can then be removed in a froth. The waste materials are left behind and can be washed away.

Roasting

Roasting removes impurities such as sulphur from the ore. The ore is heated in air and the sulphur combines with the oxygen in the air forming gases. Remaining solid material must be purified to obtain the pure metal.

Smelting

After the ores have been dressed, roasted and treated in other ways the actual work of extracting the metal can begin. Smelting involves melting the ore in such a way as to remove impurities. Iron ore, for example, is placed in a large, brick-lined **furnace**, called a blast furnace where it is fiercely heated with coke and limestone which combine with impurities in the ore and can then be removed from the furnace.

Leaching

Leaching involves dissolving the metal out of the ore using a chemical **solvent**. The metal is recovered from the **solution** by adding another chemical that causes the metal to separate out from the solution.

Some metals can be separated from their ore by a process called leaching.

Electrolysis

Some substances are so tightly bound together that they can only be torn apart using the energy of electricity. Using an **electric current** to split up a substance is called electrolysis. Electrolysis will only work with substances that **conduct** electricity and is used for refining copper and extracting aluminium, for example.

Electrolysis at work

Two solid electrical conductors, called **electrodes**, are placed into a liquid and connected by wires to the terminals of a battery or a generator. The liquid must contain a substance that can carry the current and complete the electrical circuit. This is called an electrolyte. As the current flows, chemical changes take place at the electrodes. The electrode connected to the battery's negative terminal is called the cathode. It attracts positive ions from the **solution**. The other electrode is called the anode and it attracts negatively charged ions from the solution. When solutions containing metals are electrolysed the metal is deposited on the cathode.

If rocks containing copper are dissolved in sulphuric acid a current can be passed through the solution. Copper in the solution is attracted to the cathode where it can be collected.

Electrolysis is also used to produce sodium metal by the electrolysis of molten sodium chloride (common salt). Chlorine gas is produced at the anode and sodium at the cathode. Both sodium metal and chlorine gas have many important industrial and chemical uses.

Try it yourself

You will need
salt
a glass of water
a 9-volt battery
two lengths of copper wire
aluminium foil
sticky tape

1 Cut out two small squares of foil and tape
   them to one end of each wire.
2 Fix the other ends of the wires to the battery terminals.
3 Dissolve a spoonful of salt in the glass of water and dip
   the foil squares into the glass.

Bubbles of gas will start to form on the foil
squares as the electric current splits the salt
and water molecules.

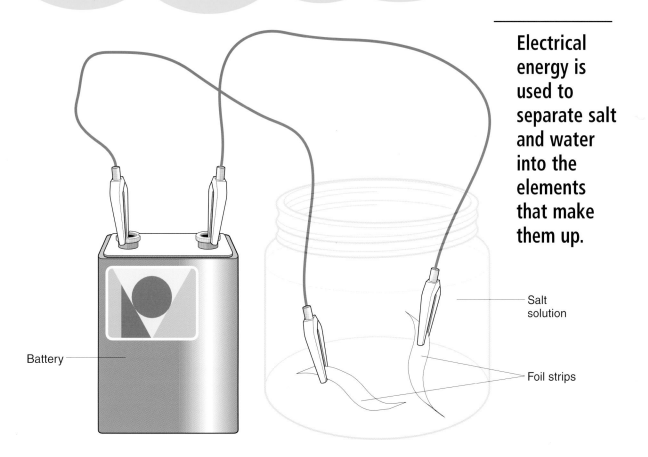

Battery

Salt
solution

Foil strips

**Electrical
energy is
used to
separate salt
and water
into the
elements
that make
them up.**

Chromatography

Chromatography is a way of separating a **mixture** of substances. The mixture is **dissolved** and then passed through something such as a **filter** paper or oil, with which it will not mix.

Different substances in the mixture are more or less strongly attached to the surface of the material they are passing through. This means that they will travel through at different rates and this makes them separate out. Scientists often use chromatography to discover what different materials contain.

Chromatography can be used to identify tiny parts of a mixture. Computers can be used to help identify an unknown mixture by looking for various properties of the possible ingredients.

Types of chromatography:

Liquid column chromatography
The mixture is dissolved and added at the top of a glass tube filled with a material through which it can travel. The substances move down through the column at different speeds, separating out as they do so.

Thin layer chromatography
A drop of the mixture is placed on one end of a thin, flat sheet of glass or other material coated with a thin layer of a substance through which the mixture can travel. The sheet is stood on end in a shallow pool of liquid which travels up the thin layer, carrying the mixture along with it.

Gas chromatography
Is used to separate gases and substances that are easily converted into gases by heating.

Try it yourself

You will need
water
ink or food colouring
a glass jar
a pencil
a bulldog clip
filter paper or blotting paper
a dropper

Bulldog clip
Pencil
Original position of dye

1 Place a little water in the bottom of the jar.
2 Attach a strip of filter paper to the pencil using the clip. The strip should be just long enough to reach the water in the jar when the pencil is laid across the top of the jar.
3 Use the dropper to place a drop of ink or food colouring on the end of the filter paper.
4 Suspend the paper in the jar with the strip just in the water. The pencil will stop it from falling in.
5 As the water carries the ink up through the paper, the different components in it will separate out as they travel through the paper at different rates.

Recycling

A great many of the things we use can be recycled. Recycling means collecting, processing and reusing materials instead of throwing them away. Paper, glass and aluminium and steel cans can all be recycled. Doing this saves **raw materials** and energy. It also saves the space that would have to be used to deposit all the unwanted rubbish and helps reduce the pollution caused by burning waste.

Recycling materials not only saves raw materials and energy, it also reduces the need for unsightly waste disposal sites.

Recycled materials can be used in a variety of products. Aluminium from recycled cans can be used to make new cans and other products. Recycled paper can be used for insulation as well as in making paper and cardboard. Waste glass can be ground up and melted to make new glass containers. Some plastics can be melted and reshaped to make new plastic products. Motor oil can be recycled and used as industrial fuel oil.

Energy efficiency

Melting down an aluminium can for recycling takes just five per cent of the energy that would be needed to make a completely new can.

Sorting out the rubbish

Materials for recycling have to be separated out from other waste materials. Your town may have bottle banks for different coloured glass, as well as containers for the collection of paper, cans, clothes and other materials. Sorting materials out in this way helps to prevent them from becoming mixed up with unwanted substances and so increases their value.

Waste-processing plants separate heavier materials such as metals, glass and plastic containers from lighter materials that can be burned to produce power for the plant. A conveyor belt carries the rubbish past **electromagnets** and various mechanical devices that separate out metals and lighter objects from heavier glass. The metals can be sold as scrap or melted down and reused.

The metals in old cars can be reused. The car body is shredded and large electromagnets used to remove the iron and steel. Other metals, such as aluminium, copper, zinc and lead, can be separated by hand.

Powerful electromagnets are used in scrapyards to separate out iron and steel from other waste.

Glossary

alloy mixture of two or more metals, or a metal and a non-metal

atoms tiny particles from which all materials are made; the smallest part of an element that can exist

carburettor device in a car engine for mixing a fine spray of fuel with air

ceramic a non-metallic solid that stays hard when heated

chemical reaction a reaction that takes place between two or more substances in which energy is given out or taken in and new substances are produced

compound a substance that is made up of atoms of two or more elements

condense to change from a gas into a liquid

conduct to transmit electricity or heat by conduction

density the compactness of a substance

dissolve to become incorporated into a liquid and form a solution

distillation a way of separating a pure liquid from a mixture

electric current a flow of electricity through a substance

electrodes conductors through which electricity enters or leaves something

electromagnet a magnet produced by running an electric current through a coil of wire

element a substance that cannot be broken down into simpler substances by chemical reactions; an element is made up of just one type of atom

evaporation process by which a liquid turns into a vapour without reaching its boiling point

filter a device for removing particles from a liquid or gas passed through it

filtrate liquid or gas that has been passed through a filter

furnace a chamber in which materials can be heated to a very high temperature

leach to remove a substance from a material by dissolving

micro-organisms living things too small to be seen with the naked eye

minerals naturally occurring solid substances; substances obtained by mining

mixture material made of different substances mixed together but not combined chemically

molecule two or more atoms combined together; if the atoms are the same it is an element, if they are different it is a compound

ore rock from which metals can be extracted

particle tiny portion of matter

pesticides substance used to destroy pests

pressure a force pushing on a given area

raw material material in its natural state

residue the solid material left behind when a liquid or gas is passed through a filter

solute the material dissolved in a solvent to form a solution

solution a mixture of one substance dissolved in another

solvent the part of a solution that dissolves the solute

suspension a liquid containing small particles of a solid

vapour a type of gas

volatile a liquid that readily turns into a vapour

volume the amount of space an object takes up

Index